Stephen Roper

Roper's Instructions and Suggestions for Engineers and Firemen

who wish to procure a license, certificate, or permit to take charge of any

class of steam-engines or boilers, stationary, locomotive, and marine

Stephen Roper

Roper's Instructions and Suggestions for Engineers and Firemen
who wish to procure a license, certificate, or permit to take charge of any class of steam-engines or boilers, stationary, locomotive, and marine

ISBN/EAN: 9783337254506

Printed in Europe, USA, Canada, Australia, Japan

Cover: Foto ©berggeist007 / pixelio.de

More available books at **www.hansebooks.com**

ROPER'S

INSTRUCTIONS AND SUGGESTIONS

FOR

ENGINEERS AND FIREMEN

WHO WISH TO PROCURE A

License, Certificate, or Permit

TO TAKE CHARGE OF

ANY CLASS OF STEAM-ENGINES OR BOILERS, STATIONARY, LOCOMOTIVE, AND MARINE.

BY

STEPHEN ROPER, ENGINEER,

AUTHOR OF

"Roper's Catechism of High-Pressure or Non-Condensing Steam-Engines,"
"Roper's Hand-Book of the Locomotive," "Roper's Hand-Book of
Land and Marine Engines," "Roper's Hand-Book of Modern
Steam Fire-Engines," "Improvements in Steam-Engines,"
"Use and Abuse of the Steam-Boiler," "Questions
and Answers for Engineers," "Care and Man-
agement of the Steam-Boiler," etc., etc.

———◆———

PHILADELPHIA:
E. CLAXTON & COMPANY,
No. 930 MARKET STREET.
1884.

INTRODUCTION.

IT is not the intention of the writer, in the preparation of this book, that it should supersede his work entitled "Questions and Answers for Engineers," or interfere with anything that he has previously written, as it is a collection of theories and practical facts which have been recently demonstrated in connection with Steam-Engine and Steam-Boiler Engineering.

All Engineers and Firemen should avail themselves of every source of information relating to the duties of their calling; for it is a fact, admitted by intelligent engineers, that the greater effort we make to acquire knowledge in relation to the steam-engine, the more thoroughly do we become convinced of how little we previously knew about the subject.

No possible or even plausible reason can be assigned why men in other pursuits should expend large sums of money, and devote years to study, for the purpose of qualifying themselves for their profession, while Engineers and Firemen, a very numerous and important class in manufacturing and steam using communities, should assume the responsibility of duties for which they have literally made no preparation whatsoever.   S. R.

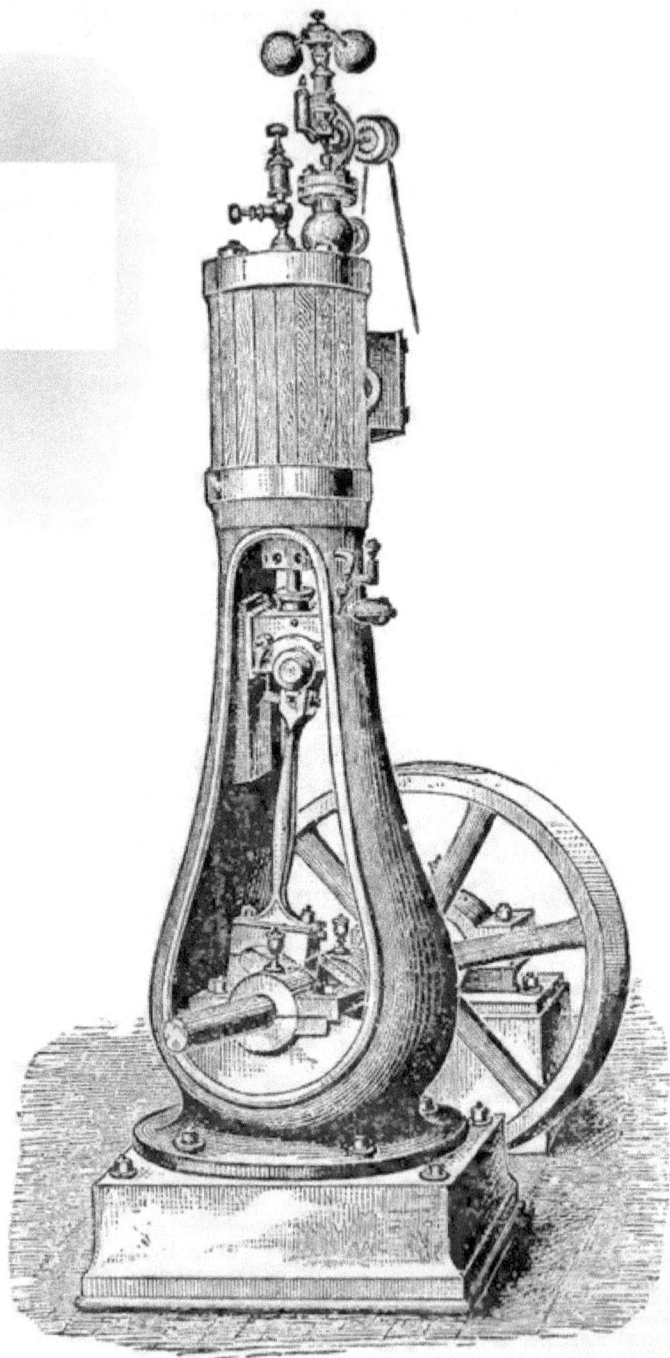

ROPER'S

INSTRUCTIONS AND SUGGESTIONS

FOR

ENGINEERS.

The first thing that should attract the attention of the engineer or fireman on entering a boiler-room in the morning is the glass water-gauge, so that he may ascertain if the water is at the proper level in the boiler.

An engineer or fireman should always try the gauge-cocks before starting his fire in the furnace, as the water may become low during the night from blowing off at the safety-valve or from leakage through auxiliary or other valves.

Every engineer or fireman in charge of a steam-boiler should blow out the water-gauge and gauge-cocks every morning, in order to remove the soft

11

mud which settles in them at night when the boiler
is at rest.   If this is neglected, the soft mud may
become baked, which might lead to disastrous results.

An engineer should close the steam and water-
valves which form a communication between the
boiler and the glass gauge every night, as, if the
glass tube should break in his absence, the escaping
steam and hot water from the boiler would most
certainly injure any property which might be in the
immediate vicinity.

An engineer or fireman on entering the boiler-
room in the morning should always ascertain whether
the cocks or valves which connect the water-gauge
with the boiler are shut or open ; otherwise he may
be deceived by the appearance of the water in the
tube.   This precaution should never be neglected.

Every engineer or fireman should blow out the
gauge-cocks regularly, not only to ascertain the height
of the water in the boiler, but to prevent them from
becoming choked with sediment or mud.

All engineers should remember that when gauge-
cocks are long and have a small orifice, they require
more attention than if the barrel of the cock was
short and the orifice large.

No engineer or fireman should allow the gauge-
cocks to leak at all when it is practicable to repair

them, as the longer they leak the more difficult they are to repair, as the material wastes rapidly under the escape of the water or steam.

No engineer should allow the gauge-cocks, glass water-gauge, or steam-gauge to become filthy, as it evidently shows want of care, and furnishes evidence that those who are not particular in this part of their duty are not reliable in others of equal or more importance.

Any engineer or fireman who values his reputation should show by the general appearance of everything in his charge that it is properly treated and cared for.

If an engineer should discover that there was too much water in the boiler, he should blow it down to the proper level; but in doing so he must exercise judgment, vigilance, and care, especially if there is a fire in the furnace.

Every engineer should be sure that all the braces, whether adjusted by swivels, turn-buckles, lugs, or cotters, meet the requirements for which they were intended, as no brace is of any value unless it is taut.

Every engineer should understand that braces, when subjected to extreme pressure, become elongated, and receive what is termed a *permanent* set. Under such conditions they offer no guarantee of safety.

If any accident should occur, such as the bursting or leakage of a pipe, the pumping of the well dry, or the choking of any of the pipes which convey the water from the source of supply to the pump, the engineer should shut down his engines, cover the fire with fresh fuel, close the damper, and keep sufficient water in the boiler until. the difficulty is overcome, or the damage repaired.

Every engineer or fireman should raise the safety-valve every morning to ascertain if it is in good order; he should also compare the indications on the steam-gauge with those of the safety-valve when it blew off, in order to satisfy himself whether they are correctly marked, or if they work in unison or not.

Every engineer should know that steam-boilers are in many respects similar to men and animals. When well cared for, kept clean, and not overtaxed, they render efficient service, but if abused or neglected they become inefficient and dangerous.

Every engineer should frequently clean the tubes and flues of the boilers in his charge, as the ashes which settle in the lower surface of the tube or flue is a non-conductor, and induces waste of fuel. He should also clean the outside of the shell crown-sheet and all accessible parts of the boiler.

An engineer should never undertake to caulk a boiler, for the purpose of preventing leakage, while it contains either water or steam, as, while he may make it tight in one place, he will be sure to cause it to spring in another.

An engineer should remove the ashes from under the boilers as often as practicable, as such accumulations retard the draught and interfere with combustion, thereby causing waste of fuel, and interfering with the evaporative efficiency of the boiler or boilers.

When it becomes necessary to blow down the boiler or boilers at intervals, the engineer or fireman should stand by the blow-off cock, and not allow his attention to be diverted to anything else, as in a very short space of time the water may become too low, induce stoppage, or endanger the safety of the boiler.

No engineer should disturb the safety-valve when there is a high pressure of steam in the boiler and a heavy fire in the furnace, as it would set in motion a large volume of steam of a high temperature and great elastic force, which might blow off the safety-valve, or result in the destruction of the boiler.

Engineers and firemen should always be cautious when they either stop or start an engine with a heavy pressure of steam in the boiler, as the vent given to the steam when starting, and the check which it receives in stopping, may exert such a pressure as to strain, crack, or rupture the boiler.

An engine should be started slowly, and with just sufficient momentum to carry the crank over the centre, and then bring it gradually up to its regular speed, as, when started in haste, the water which results from the condensation of the steam when it comes in contact with the cold cylinder is liable to fracture the piston or cylinder, or spring the joints, and cause them to leak.

An engineer should always admit a small quantity of steam to the cylinder, and then move the crosshead back and forth on the guides, for the purpose of heating up the cylinder, and expelling the water of condensation.

An engineer should leave the drip-cocks in the cylinder open whenever the engine is standing still ; they should not be closed until after the engine has been started up and made several strokes or revolutions.

No engineer should ever open the stop-valve to its full extent on starting up after the engine has been standing still over night, as the quantity of steam condensed by being brought in contact with the cold pipe, particularly if it is long, may result in breaking the follower-plate, springing the piston, or knocking out the cylinder-head.

Every engineer should be aware that all valves, whether connected with boilers or engines, leak after they are a short time in use; this period depending on the quality of the metal of which they are made, and the character of the workmanship employed in fitting them up.

Every engineer should understand that, though the process of leakage may go on slowly through the valves and stop-cocks connected with the boilers in his charge, they nevertheless go on surely and steadily; and, as constant dropping wears the stone, so does constant leakage at night deplete the water in the boiler and the fuel in the daytime.

An engineer should grind in the steam, water, safety, and auxiliary valves and stop-cocks as often as their condition may require, as every gallon of water lost by leakage of steam requires the consumption of so many pounds of fuel.

Every engineer should understand that though the glass water-gauge is a very convenient arrangement, it is not as reliable as the gauge-cocks, because the very process of using them serves to keep them in good order.

When an engineer or fireman opens the gauge-cocks to ascertain the height of the water in the boiler, he should close them tightly to prevent leakage.

Most engineers have discovered, when gauge-cocks are closed after being blown out, they leak badly,

2 *   B

which is often due to the fact that some mud or sand has become attached to the seat of the valve or the point of the plug. The easiest way to remedy this difficulty is to open the cocks and let them blow out for some time, when the friction of the water, in its escape through the orifice, will in all probability remove the obstacle.

Every engineer should understand that glass gauges may be cleaned by closing the water-valve, opening the drip-cocks, and allowing the steam to blow through. This will have the effect of detaching the mud from the inside of the glass.

Every engineer should know that glass water-gauges may be cleaned by tying a piece of cotton-waste or lamp-wick to the end of a rattan or a splint of wood, applying some soap or acetic acid to it, and passing it down the inside of the tube. Then, by opening the drip-cock, the glass may be washed out, and appear as bright as when new.

An engineer should never touch the inside of the glass water-gauge with iron or wire, as, while the gauge may be cut on the outside with a file, the slightest touch of iron or steel on the inside will cause an abrasion, the result of which is that the glass cracks and becomes useless.

Engineers, when purchasing glass gauges, should understand that those of American manufacture are not reliable; that the Scotch gauge termed the Eureka is the only reliable gauge in use. It is of a light-greenish color, and displays a fibre nearly like that of cloth. The Eureka gauge commands the market of the world, as it does not appear that the secret connected with its manufacture has ever been discovered.

Engineers should know that glass gauges frequently break because the steam and water connections are not in line, because the stuffing-boxes are screwed down too tight, and sometimes, in cold weather, when a draught of cold air is admitted by the opening of a door or window. They generally give way near the water connection, rarely near the steam-valve.

An engineer should never blow out a boiler under a high pressure of steam, as the change of temperature has a tendency to cause sudden contraction, which is liable to induce leakage, and, in many instances, fracture of the rivets, braces, or shells, where there are sharp bends, as in the corners of furnaces, flanges, flues, and tubes, etc.

An engineer should always blow out his boiler under a moderate pressure, say thirty pounds per square inch, after which the boiler should be allowed to cool off gradually before refilling.

No engineer should ever fill a boiler with cold water while it is hot, as the injurious effect produced by contraction is similar to that induced by blowing out under high pressure, and, if persisted in, will result in permanent injury to the boiler.

An engineer should never allow the tubes, flues, or crown-sheet of a boiler to become dry before cleaning, as the heat given out by the brick-work and the different parts of the boiler has a tendency to attach the scale firmly to every part with which it comes in contact, rendering its removal almost impossible.

Every engineer, after blowing out his boiler preparatory to cleaning it, should fill it again with cold water, and let it remain until he is ready to commence operations. The mud will then be in a soft state, and can be easily scraped, swept, or washed out.

An engineer, before blowing out his boiler, should remove all the fire from the furnace, as a small quantity left in the corners, or attached to the bridge-wall, might spring a seam or cause a plate to bulge.

Every engineer, when starting a fresh fire under a cold boiler, should allow it to burn moderately, in order that all the parts may be heated gradually and expand uniformly, otherwise some parts of the boiler will be heated to a high temperature, while others will remain nearly cold, which will have an injurious effect on the boiler.

Every engineer should know that unequal expansion and contraction is one of the evils which limit the longevity and endanger the safety of all classes of steam-boilers, consequently, in blowing out, the refilling, the starting of fires, and the regulation of the draught, should be done with judgment.

Every engineer should know that all the care and precaution exercised in the management of steam-boilers brings its reward either in durability or economy, obviates deterioration, danger, stoppage, delay, repairs, cost of maintenance, and eventually destruction, perhaps accompanied by disaster and loss of life and property.

Every engineer should be capable of discriminating between the different shaped boiler-heads most generally in use, whether made of wrought- or cast-iron. They may be designated as follows: the " flat-head," the " concave," the " convex," or " egg-end."

Every engineer should know that the flat design for a boiler-head is the worst disposition that can be made of a certain amount of material to insure strength and safety, especially when cast-iron is employed.

Engineers should understand that the concave is the strongest form of boiler-head, but it has the disadvantage that it occupies more space in the end of the shell than any other design, and, if placed directly over the fire, the shell and rivets will burn off, as the water cannot come in contact with them, owing to the great distance which the head protrudes into the shell. To obviate this, it is customary to place the head outside of the furnace, or over-hang it, which proves an element of waste, as the more a steam-boiler is exposed to the atmosphere, the more fuel will be required to evaporate a certain quantity of water.

Engineers should fully understand the mechanical principles involved in the design of different boiler-heads. The flat-head may be said to be a lever, which, when pressure is exerted against it, has a tendency to bulge out and crack the material at the point where the flange is formed, which connects the head to the shell.

Every engineer should know that the reason why boilers are double-riveted on the longitudinal seams and single-riveted on the curvilinear seams is, that there is twice the pressure on the former as there is on the latter.

Every engineer should understand that, in making calculations on the strength of boilers, we should employ the factor 56, instead of 100, as 44 per cent. of the strength of the plate is removed in punching the holes for the rivets; the remainder must form the basis of the calculation.

Every engineer should understand that single-riveted seams are equal to 56 per cent. of the original strength; that double-riveted seams are equal to 70 per cent., and that triple-riveted seams are equal to 85 per cent., etc.; but it must be understood that triple-riveted seams are very seldom used, unless for some special purpose, as they are too heavy and thick, and would burn out rapidly if exposed to the fire.

Every engineer should understand that machine-riveted seams in steam-boilers are superior to hand-made ones, as the machine thoroughly upsets the rivet, and brings the two sheets in such close contact as to produce friction between the sheets at the lap, which of itself is an element of strength.

Every engineer should understand that the shearing strain on hand-riveted seams is greater than on those riveted by machinery, as the cohesion of the lap is much less in the former than in the latter method.

All engineers should understand that counter-sunk rivets do not possess the same strength as flush-heads of the same diameter and space apart.

Every engineer should understand that the strains to which boiler-heads are exposed, when subjected to pressure, are either bending, as in the case of the *flat head*, crushing, as in the *concave head*, or tensile, as in the *convex head*, and that the power of any boiler-head to resist pressure depends on the quantity and quality of the material of which it is made and the sectional area between the rivets by which it is attached to the shell.

All engineers should understand that boiler-heads of any considerable diameter require to be braced both with angle-irons and diagonal braces to insure safety.

Every engineer should understand that the domes of steam-boilers should be braced down to the crown-bars and the crown-bars braced up to the dome. This is an imperative necessity, particularly in the case of marine and locomotive boilers.

Engineers should understand that braces in steam-boilers are only a substitute for strength, because, if the boiler possessed sufficient strength to resist the pressure, there would be no need of braces.

All intelligent engineers can easily understand that the value of any brace depends on the angle at which it has to sustain the strain, push, or pull; if the pulls be straight, the resistance of it before giving way will depend on the tensile strength of the material of which it is made.

Every engineer should understand that the ends of braces should never be tapped into the shells of boilers, except in the fire-box, water-legs, or at the junction of the waist; they should in all cases be connected with lugs riveted to the shell, for which purpose two rivets of moderate size are preferable to one large one.

When an engineer discovers that the braces in the boiler under his charge are slack, he should have them taken out and upset. This may be easily done by heating the brace near its centre in a blacksmith forge, and then striking it on the end with a mall or wooden mallet, or by striking it against a block of wood while it is hot. Cotters sometimes become worn on one side. In such cases they should be turned round or taken out and replaced with new ones.

Every intelligent engineer should know that the elasticity of the shells and flues of steam-boilers is so limited that, unless the braces are all taut, the parts of the boiler which they were intended to strengthen will be in danger before the brace sustains any portion of the strain.

Every engineer should understand that, if the shells of boilers possess sufficient strength to resist the pressure, there would be no need of braces. Nevertheless, it is impracticable in many cases to dispense with them, as in the case of marine boilers of very large diameter.

Some engineers may think that boilers might be constructed in such a manner as to dispense with the necessity of braces. This is evidently a mistake, which will be shown in the following paragraph.

An engineer should understand that the value of a brace depends, to a certain extent, on the methods by which it was attached to the boiler. The resistance of a brace in the direction of the push will depend on the following two factors, viz., whether

3

the weight is transmitted in a direct line or not, and on the liability of the brace to bend or stretch out.

Every engineer should understand the use of all the braces in the boiler under his charge ; the direction in which they sustain the strain, the amount of strain they are capable of sustaining with safety, and the arrangements by which they are attached to the different parts of the boiler, etc.

Every engineer should be familiar with the names of the different braces, which may be designated as follows : *water-leg braces,* which are in turn termed tap bolts, front braces, back braces, dome braces, crown braces, swivel braces, angle braces, toggle braces, crow-feet braces, cotters, etc.

Every engineer should know why boiler braces are designated by the foregoing names. This may be explained in this manner : angle braces are generally employed for the tube-sheets of boilers, in order to give additional strength above the tubes in the steam room. They are rarely ever used in the water-space. The dome brace is located between the crown-bars and the dome, for the purpose of resisting the upper pressure against the dome and the downward pressure on the crown-sheet. They are riveted into the dome and straddle the crown-bars. Toggle braces are either riveted in at both ends or are attached to the shell by lugs. They have their adjustment in the centre, which consists of a swivel or turn-buckle. Diagonal braces are the weakest of all.

Every engineer should understand that, to make a boiler of large diameter of sufficient strength to resist any pressure to which it might be subjected without the aid of braces, the plate should be extra heavy, and triple-riveted, which would more than double the first cost of the boiler. Such thick plates and such enormous seams would soon burn out.

Engineers should understand that moderate diameters and thickness of plate, single-riveted seams for the curvilinear and double for the longitudinal seam, give the best results.

Every engineer should understand that many conditions influence the strength, efficiency, economy, durability, and safety of steam-boilers.

It should be well understood by engineers that, if boiler-plate is made of refined iron, and sufficient thickness to give a reasonable margin of safety, it will give economical results.

Every engineer should know the formulæ for finding the weight on the safety-valve, or the weight to be placed on it to carry a certain pressure, which reads as follows:

Multiply the area of valve by the pressure in pounds per square inch; multiply this product by the distance of the valve from the fulcrum; multiply the weight of the lever by one-half its length (or its centre of gravity); then multiply the weight of the valve and stem by their distance from the fulcrum; add these last two products together, subtract their sum from the first product, and divide the remainder by the length of the lever; the quotient will be the *weight* required.

Every engineer should be able to repeat from memory the formulæ for finding the safe working pressure of any boiler, which may be expressed thus:

Multiply the thickness of the iron by 56, multiply this product by 10,000 safe load, divide by the external radius less the thickness of the iron; the quotient will be the safe working pressure in pounds per square inch.

Every engineer should be able to find the amount of heating-surface in any boiler, which may be ascertained by the following formulæ:

Multiply two-thirds the circumference of the boiler in inches by its length in inches; multiply the area of one flue or tube in inches by its length in inches; add all these products together and divide by 144; the quotient will be the square feet of heating-surface.

Every engineer should observe the indications on the steam-gauge when the safety-valve blows off, and if the gauge indicates several pounds per square inch ahead of the valve, or lags behind it, it is evident that either the valve is marked wrong or the gauge is out of order.

An engineer should have the steam-gauge attached to the boiler under his charge tested every year, in order to ascertain if it is correct in its indications. This can be done at any establishment where steam-gauges are made.

Every engineer should understand that the valves and stop-cocks connected with the boiler in his charge should be ground-in frequently, that they should never be allowed to leak, and that the best material for grinding valves or stop-cocks is pulverized glass. It is much superior to emery.

Every engineer should clean out the boiler in his charge at least twice a year, or oftener, if practicable.

Every engineer should understand the rule for estimating the quantity of water that any boiler will hold, either in cubic feet or gallons, which is as follows:

Multiply the area of the head or end in inches by the length of the shell in inches, and divide by 1728; this will give the cubic feet of water from which the whole number of flues or tubes must be subtracted. The remainder, if multiplied by 7.5, will give the quantity in U. S. gallons.

Every engineer should understand that the term heating-surface means every part of the shell, tubes, flues, tube-sheets, crown-sheet, and furnace with which the hot gases come in contact in their escape from the furnace to the chimney.

3 *

All engineers should know that steam-boilers are stronger when under a given steam-pressure than when under the same hydraulic or cold-water pressure, because iron toughens by the application of heat up to 500° Fah., but any increase of temperature above that degree weakens it.

Every engineer should understand that the efficiency of steam-boilers is influenced by many conditions, such as design, quality of material, draught, location, fuel, and management. If the design insures good circulation, if the boiler-plates are of high-conducting materials, the fuel of good quality, the draught good, the location desirable, and the attendance and management careful and intelligent, the result will be satisfactory and economical.

Every engineer should understand that scale and incrustation is a non-conductor, and induces waste of fuel, which varies, according to the thickness of the scale, from 1 to 37 per cent.; the scale also prevents the water from coming in contact with the boiler-plate, and induces crystallization and brittleness, which renders it liable to crack, bulge, or blister.

Every engineer should understand the difference between hard and soft patches. In the former, the patch is attached to the part by means of rivets, and caulked in the ordinary way, rendering it what is termed iron and iron or solid; while in the case of the *soft patch* the defect is repaired by fitting a patch to the hole or crack, then applying a packing of white lead, litharge, or putty between the two surfaces, and tying them down with bolts, nuts, and washers. Under such circumstances the holes in both the patch and the part of the boiler to which it is attached are often of larger diameter than the bolts, which circumstance allows the patch to move; besides, the packing is not positive, and, though it may make a steam- and water-tight joint for the time, instead of being a guarantee of safety, is an element of danger.

Every engineer should understand that a riveted flue will not stand as much pressure as a rolled tube of the same diameter and of less thickness. This is due to the fact that riveted flues are not perfectly round, and any deviation from a true circle is liable to cause collapse.

Every engineer should understand that a boiler-flue of a given diameter would stand twice as much external pressure before collapsing as a flue of the same diameter and twice the length.

No engineer should ever fill a boiler with cold water above the second gauge-cock, as any more is not necessary, as water swells under the process of a formation of steam, and it will be found that there is a sufficiency of water in the boiler when steam is raised.

Every engineer should understand that water vaporizes or becomes steam at 212° Fah., and that it freezes at 32° Fah., and attains its greatest density at 39° Fah.

Every engineer should know that water is almost incompressible; that, when confined, it is nearly as solid as iron, and that the increase in volume between water and steam is 1700. One cubic foot of water converted into steam at atmospheric pressure will make 1700 cubic feet of steam.

Every engineer should understand that boiler explosions are the result of weakness either in the shell, flues, heads, braces, or other parts of the boiler. This weakness may be due to poor material, inferior workmanship, over-firing, incrustation, corrosion, cracks, flaws, etc.

Every engineer should understand that there is no mystery about boiler explosions. They are all cause and effect, and when a boiler does explode, it is certain that it gave way in the weakest part, and that the pressure was too strong for the boiler, or that the boiler did not possess sufficient strength to resist the pressure.

No engineer should ever entertain the vagaries that are frequently advanced in regard to boiler explosions; among which are urged the gas, electricity, and other abstruse theories, as experience has demonstrated that all such ideas are chimerical.

Every engineer should understand that an explosion caused by an insufficiency of water is less destructive in its effects than if the same boiler exploded when it contained a sufficient quantity of water, because, when a boiler explodes from want of water, certain parts become overheated and softened, and it does not require as much pressure to burst it as if it contained the necessary amount.

Every engineer should understand the meaning of the term collapse, which is the flattening of a *fluid* or tube. Accidents resulting from collapse are becoming rare since the rolled tube has to a certain extent superseded the riveted flue.

Every engineer should understand that the term shearing of the rivets means cutting them off by pressure on the tube-plates which form the riveted seam. The sectional area between the holes in drilled seams is stronger than if the holes were punched, because the punch disturbs a certain portion of the material around the outside of each hole, while the drill does not, and therefore boiler-seams in which the holes are drilled are more liable to shear than if they were punched, on account of the sharpness of the edges of the drilled holes.

C

What is a Compound-engine?—A two-cylinder engine in which the steam is used twice.

Explain the principle embodied in the working of the compound-engine.—A compound-engine has two cylinders, one high- and the other low-pressure. The steam is admitted from the boiler to the high-pressure cylinder, is cut off at half-stroke, then escapes to the low-pressure cylinder and follows the piston ⅞ of the stroke, after which it is released, and enters the condenser.

Are Compound-engines invariably condensing? No; some compound-engines have two low-pressure cylinders and no high-pressure, while others have two high-pressure cylinders and no low-pressure; the majority of them have one high- and one low-pressure cylinder, but in all cases the steam is used twice.

What are the advantages and disadvantages of the compound-engine?—The advantages claimed for the compound-engine are that it affords a better distribution of the strains due to expansion and contraction than the simple engine, and that it admits of a greater degree of expansion than the simple engine.

How much of the heat stored up in good fuel is utilized in the best class of steam-engines?—About ten per cent. in non-condensing engines, and not more than twenty in condensing engines.

How can you explain the foregoing assertion?— If we place a thermometer in the steam-pipe, between the boiler and the cylinder, it will register the temperature of the incoming steam. Now, if we place another in the exhaust-pipe, it will show that the steam has not lost much of its temperature, consequently the difference in temperature, or the heat lost by the steam while passing through the cylinder, was all the benefit derived from the consumption of the fuel.

What course would you pursue in case the Eccentric slipped or turned round on the shaft and stopped the engine?—I would place the crank on the centre, remove the bonnet of the steam-chest, and move the eccentric round in the direction in which it used to run until the valve had the proper amount of lead, then I would tie it down with the set-screw or key, as the case may be.

What course would you pursue in case the stop-valve stuck on its seat, and could not be opened?—I would slacken the nuts which attach the bonnet to the body of the valve, which would have the effect of lengthening the space between the face of the valve and the screw on the spindle, and admit of opening the valve without difficulty.

Why should valves stick on their seats?—Because they are frequently shut when cold, and when heated up by the steam the valve-stem becomes lengthened.

What course would you pursue if the Stem of the Stop-valve broke off inside of the Stuffing-box?—I would shut the auxiliary-valve, take off the bonnet of the broken valve, and open it sufficiently to admit the necessary quantity of steam, after which I would replace it, and shut off or start the engine by means of the auxiliary-valve.

If there were no Auxiliary-valve on the steam-pipe, how would you stop or start the engine if the main valve was broken?—I would throw the eccentric-hook out of gear, and stop the engine by means of the starting-bar.

In case you undertook to screw up a nut on a bolt connected with the boiler, the steam-chest, or cylinder of the engine, and it broke off and allowed the steam to escape in great volume, what would you do?—I would drive in a wooden plug with a sledge or heavy hammer, or I would cover the hole with a piece of gum packing and an iron plate, and brace off with a piece of scantling to a wall or post; the brace should be at least one inch longer than the space.

In case the Eccentric should become worn flat at two points in the direction of the push and pull, what course would you adopt to remedy the difficulty?—I would file it down to the smallest diameter of the worn part.

What is the meaning of the term Throw of the Eccentric?—It means the distance that the eccentric is out of centre.

4

How can you find the Throw of the Eccentric ?—Measure the light and heavy sides, and the difference between them is the throw.

What should be the Throw of Eccentrics for Slide-valve Engines? — Double the width of the steam-ports and the *lap* added.

Suppose the Cross-head and Wrist-pin boxes become brass and brass or brass-bound, so that the lost motion cannot be taken up, what course would you adopt?—I would have them filed off on their top and bottom edges.

Where would you look for Knocks in the Steam-engine?—In the connecting-rod boxes and the piston, in the valve, in the follower-plate, in the key which connects the piston-rod to the cross-head, in the pillow-block box, and in the fly-wheel.

What is the cause of Knocking in Steam-engines? —It is due to looseness in the boxes or other parts of the engine, or because it is out of line.

What is a Thud?—A thud is a peculiar noise made by an engine when the valves are not properly set, and the engine does not take her steam and let it go at the right time.

What is the meaning of the term *"Lap"* on the Valve?—The amount that the valve overlaps both steam-ports when the valve is in the centre of its travel.

What is the meaning of the term *"Lead"* on the Valve?—The amount of opening the valve has when the crank is at the commencement of the stroke.

What is the meaning of the term Steam and Exhaust *Lead?*—Steam *"lead"* means the amount of opening the valve has on the steam end when the crank is on the centre. Exhaust *"lead"* means the amount of opening the exhaust has when the crank is at the same point.

Providing a Gauge-cock becomes broken off near the boiler-head, what course would you pursue?—I would plug up the hole with a piece of wood, and use the other two.

In case the glass water-gauge should break, and discharge large quantities of hot water and steam into the room, what course would you pursue?—I would place my hand in an old felt hat or cap, or hold up a coat before my face, and shut off the water-valve first, and then the steam, and run by the gauge-cocks.

If you discovered that the cylinder was worn hollow in the middle, what course would you adopt? —I would have it bored out.

Suppose the slide-valve became leaky from wear, how would you treat it?—I would take it out, have it planed on the face, and the seat filed and scraped.

In case the Crank or Wrist-pin should be cut or

worn oval, how would you treat it?—I would caliper them and file them round.

Suppose the shoes in the Cross-head guides should become worn, what method would you adopt?—I would put in liners between the back of the shoe and the jaw of the cross-head, or replace them with a new set.

Which material do you consider the best for Cross-head shoes?—Wood is very much superior to brass, iron, or composition, as it requires very little lubrication and does not cut the guide.

What is the meaning of the term Travel of the Valve?—The distance the valve moves on its seat at each stroke of the eccentric.

How high does a Poppet-valve have to lift to give the proper opening?—One-quarter of its diameter.

What is meant by a Rotary-valve?—A valve that revolves and admits the steam, and releases it at certain points of the stroke.

What is a Semirotary-valve?—A valve that vibrates or rocks the same as a Corliss valve.

What is a Gridiron-valve?—A valve with several small openings in it.

What is a Basket-valve?—It is a valve that works within a shell, which is perforated with numerous small orifices.

Which is the most simple and durable valve?—

A slide-valve, because its first cost is very trifling, it is easy to repair or replace, and it requires no special tools for that purpose; besides, it is the most positive and reliable of all valves, as it can be run at almost any speed.

What are the objections to the Slide-valve?—The great amount of power expended in working it; besides, it is very wasteful when the ports are long.

What is the meaning of the term Admission?— Admission takes place when the valve opens to admit the steam to the cylinder, which in all cases occurs when the crank is at the centre.

What is the meaning of the term Cut-Off?—It means that the valve has closed and cut off the steam at a certain point in the cylinder.

What is the meaning of the term Compression?— Compression means that the steam in the cylinder has been gathered up by the movement of the piston and compressed as the piston approaches the end of the stroke.

What is the meaning of the term Release?—Exhaust.

What is the meaning of the terms Induction and Eduction?—They are obsolete terms, which were formerly used for admission and release.

What is Heat?—It is a species of motion, because heat produces motion, and, *vice versa*, motion produces heat.

4 *

What is Latent Heat?—Heat that is not observable by the thermometer.

What is the meaning of Fahrenheit, Centigrade, and Reaumur?—The names of the three persons who invented the three thermometers in most general use.

What is the meaning of the term Pirometer?—A pirometer is an instrument used for showing the degrees of heat in any substance that a thermometer will not record.

How many Centres are there in a revolution?—Two; the end-board and the out-board.

How can the exact dead centre of an engine be found?—By placing a spirit-level on the top side or hold it under the bottom side of the strap, on the stub-end box, and moving the crank up and down until the level is shown.

What is the meaning of the term Stub-end boxes?—The boxes which connect the connecting-rod with the cross-head, wrist, and crank-pin. They are sometimes called connecting-rod boxes, also brasses.

Can an engine be run without a balance-wheel?—Yes; a double-cylinder engine, the same as a locomotive or a marine engine, as one crank is at right angles or half-stroke when the other is at the centre; but in the case of single-cylinder engines it is necessary to have a balance-wheel to carry the crank over the centre.

Does it make any difference whether the centre of the crank-pin is above or below the centre of the guides or cylinder?—No; providing the crank-shaft is level and the centre of the crank-pin at right angles with the centre of the cylinder.

When is the crank at half-stroke?—When it stands at right angles with the cylinder.

If an engine is out of line, where does the greatest strain come?—When the piston is in its extreme length in the cylinder.

Why should there be more strain when the piston was at the the end-board centre than at the out-board? —Because on the latter the connecting-rod and crank-pin connections have a chance to spring and relieve the strain, while in the former case they do not, as the piston-head is in the back end of the cylinder, the rod held fast by the stuffing-box, and the cross-head controlled by the guides.

What are the four terms applied to the admission and escape of the steam to and from the cylinder?— Admission, release, cut-off, and compression.

What plan would you adopt to increase the power of a steam-engine?—The most practical way of increasing the power of an engine is either to put on a new cylinder, which involves the necessity of a new piston, steam-chest, and valve-rod, or I would raise the pressure in the boiler, if it was considered safe, or increase the speed of the engine.

How much would you increase the diameter of the cylinder if you wanted to increase the power of the engine?—Not over two inches, as, with any increase over that amount, the other parts of the running gear, the crank connecting-rod, cross-head, and eccentric-rod would be light.

How would you proceed to set the valves of a steam-engine?—I would place the crank at the dead centre, adjust the valve-gear so that the valves would have the right *lead* on that end; then I would place the crank on the other centre, and if the lead was just the same, the valve ought to travel properly, providing it was well designed.

What is the meaning of the term Ports?—The orifices from which the steam enters and escapes from the cylinder.

What is the proper diameter for a steam-pipe? —One-quarter the diameter of the cylinder, and the exhaust-pipe should be about one-third.

What is the proper diameter for the crank-pin of any engine?—One-quarter the diameter of the cylinder; the inside diameter of the steam-pipe and the outside diameter of the crank-pin and cross-head wrist might be the same.

Why are engines called Square engines?—Because the stroke is twice the diameter of the cylinder; for instance, 12-inch cylinder 24-inch stroke.

Why are engines called Heat engines?—Because they are worked by steam.

What are Geared, Trunk, and Oscillating engines?—The geared engine is so arranged that the propeller runs faster than the engine; the trunk engine has no piston-rod, as the connecting-rod is attached to the piston-head and to the crank-pin; while the oscillating engine has no connecting-rod, the piston-rod being attached to the crank, and the cylinder vibrates on trunnions, through which the steam enters and escapes.

How would you proceed to put an engine in line? —I would remove both heads of the cylinder, the piston, cross-head, and connecting-rod, and draw a line through the centre of the cylinder; then, if the cylinder was at right angles with the flank, at the end-board and out-board centres, the engine must be in line; if not, the variation must be remedied by moving either the cylinder or one of the pillar-blocks.

How would you proceed to locate a steam-engine? —I would first decide on the position it was intended to occupy, after which I would take the line of the building or the main shaft, if there was any, and set the engine accordingly.

What should be the capacity of a Pump or Injector for any steam-engine?—It should be capable of delivering or discharging one cubic foot of water per horse-power per hour.

How high should the valves of a pump rise to admit the necessary quantity of water?—One-quarter the diameter of the suction-pipe.

What is the composition of Incrustation or Scale? —It is composed of different minerals, such as lime, copper, iron, and sulphur, which are held in solution in the water, but which are separated from it in the process of evaporation.

What is the meaning of the terms Calorimeter and Vent?—Calorimeter means the heat in a furnace or flue, while vent means the opening between the smoke-box and the chimney.

How high will a pump lift water?—Thirty-three feet.

How far will a pump draw water on a level?—A thousand feet, if the pump and connections are perfectly tight and in good order.

Why will not a pump lift water over thirty-three feet?—Because that is the weight of the atmosphere; forty-five miles of air, thirty-three feet of water, and thirty inches of mercury form a balance.

Does a pump suck the water up from the well or river?—No; there is no such thing as suction. The plunger of the pump simply expels the air from the barrel or cylinder, and the water will follow the piston thirty-three feet.

What is a Pump?—A mechanical arrangement or device for drawing, raising, or forcing water.

What is a Steam-pump?—A small steam-engine with a pump attached to the outer end of its piston.

Are all pumps constructed in the same manner? —No; there are Lift and Force, Solid-piston, and Bucket-pumps, though some perform the functions of both lifting and forcing.

What is the cause of Knocking in Pumps?—Lost motion, leakage in the pipes, the valves being held up from their seats by some foreign substance, such as straw, shavings, or grit.

Will a pump lift hot water?—No; if the temperature of the water is high, the supply should in all cases be above the pump.

What is the object of an Air-vessel on a pump? —To relieve the pressure and prevent knocking.

To what part of a pump should an air-vessel be attached?—To the delivery-pipe; though some pumps have two air-vessels—one on the suction and the other on the delivery.

What should be the Capacity of an air-vessel?— Five times that of the pump-barrel.

What is Corrosion?—A mysterious wasting of the material of which steam-boilers are composed. Pitting and Bleeding are due to the same mysterious causes.

What is Water, Ice, and Steam?—Water is a fluid, steam is a vapor, and ice is a solid.

What is Water composed of?—Two gases, Hydrogen and Oxygen, in proportion—one of oxygen and eight of hydrogen.

How would you proceed to remove Scale or Deposit from a Steam-boiler?—I would use a pick, scraper, hose, and broom, where it was practicable to do so; but in the case of flue, tubular, and patent boilers, which cannot be entered, such operations are impossible, consequently the use of solvents has to be resorted to.

Does water exert great force by expansion while being converted into ice?—Yes; one of the greatest in nature, except the expansion of metals.

Does water increase in bulk under the influence of heat?—Yes, up to a certain temperature.

Is water compressible?—Only to a very limited extent, as, when it is confined, it is very nearly as solid as iron.

Is water an Absorbent?—Yes; water is the greatest absorbent of all fluids or metals.

What is the best water for Steam-boilers?—Rain-water.

Does the distilled water obviate incrustation?—Yes; but it induces other evils which are nearly as detrimental to steam-boilers as incrustation.

What is Saturation?—Saturation means that the water is impregnated with salt.

What is Supersaturation ? — Supersaturation means that the water in the boiler contains more than four ounces of salt to the gallon.

What are the most practical means of preventing supersaturation?—Blowing out, or the employment of the surface-condenser.

What kind of an engine is the Marine engine ?—An engine designed to occupy a certain space in a steamship, tug, or ferry-boat.

Are all Marine engines condensing-engines?—Not necessarily so. They may be either condensing or non-condensing engines.

What is the meaning of the terms High and Low-pressure engines ?—These terms have no definite meaning, as all steam-engines are either condensing or non-condensing.

Why is an engine called a condensing engine?—Because the steam escapes from the cylinder to a condenser, and is condensed into water.

What is a Non-condensing engine ?—An engine in which the exhaust steam is discharged into the atmosphere, or under atmospheric pressure.

Has the Condensing-engine any advantages over the non-condensing engine?—Yes; the condensing-engine will develop as much power with 35 pounds pressure per square inch as the non-condensing will with 50 pounds.

What is a Vacuum?— An empty space, where there is neither water, steam, nor air.

How does a vacuum affect the working of an engine?—It increases the power of the engine from 20 to 25 per cent.

Is a Condensing-engine more economical than the non-condensing engine?—Yes; it induces an economy of fuel of from 20 to 30 per cent.

Is the first cost of the condensing-engine more than that of the non-condensing engine?—Yes; it is nearly double.

Why should the cost of the Condensing-engine be so much greater than that of the non-condensing? —Because in connection with the condensing-engine it is necessary to have a condenser, a circulating-pump, an air-pump, a hot-well, etc., which costs nearly as much as the engine itself.

How is the Vacuum produced in the condensing-engine?—By opening the shifting-valve in the condenser, and allowing the steam to blow through from the cylinder for the purpose of expelling the air, then by closing the shifting-valve and admitting the injection-water to the condenser, the vacuum is formed.

What is the average Vacuum in good practice with the best modern condensing-engines?— Ten pounds.

Whether is a Jet- or Surface-condenser capable of producing the more perfect vacuum?—The surface-condenser.

Is the Jet-condenser in very general use?—No; it is being superseded by the surface-condenser, except on lakes and rivers where fresh water is attainable.

What is the effect of salt water on Steam-boilers?—It has a tendency to deposit a heavy scale, which causes the boiler to burn out, crystallize, crack, or bulge.

What is the effect of distilled water on steam-boilers?—It induces internal corrosion, pitting, and bleeding.

What is the cause of pitting, bleeding, and wasting, in steam-boilers?—Mechanical science has never been able to discover the real cause of these mysterious phenomena.

How can a Non-condensing engine be converted into a condensing-engine?—By attaching a condenser and air-pump to it.

How would you convert a condensing-engine into a non-condensing?—By breaking the connection between the engine and the condenser, and allowing the steam to escape into the atmosphere.

What is Lost Motion?—Lost motion is looseness in the sliding, revolving, or reciprocating parts of steam-engines.

How would you proceed to take up lost motion in the steam-engine?—By the gib and strap, or other suitable mechanical appliances.

What is the cause of Lost Motion?—Abrasion which results from the contact of revolving or rubbing substances.

What is the best prevention of Abrasion?—Lubrication or oiling.

How is a Vacuum maintained in the condenser?—By the air-pump and the injection-water.

Can a Condensing-engine be worked without an air-pump?—No; because while the steam may be condensed, the air which passes in with the steam will soon occupy the condenser and choke the engine.

What are the two terms applied to the water which results from the condensation of the steam and the water introduced to condense it?—The former is termed the water of condensation, while the latter is termed the injection-water.

How much water does it require to condense steam?—About twenty-six times the quantity from which the steam was formed; but this varies according to temperature and pressure.

Does the Injection-water mix with the water of condensation in the condenser?—In the jet-condenser it does, but in the surface-condenser it does not.

How is the Injection-water and the water of condensation extracted from the condenser?—It is drawn out from the channel-way through the foot-valve by the air-pump and delivered into the hot-well.

How is the Injection-water introduced into the condenser?—In the jet-condenser it rises through the ship's side, and enters the condenser through a rose similar to the nozzle of a garden watering-pot, while in the surface-condenser it is lifted from the sea or river by the circulating-pump, forced through the condenser, and is discharged overboard.

Can a perfect Vacuum be attained?—No; and if it was possible, it could not be maintained, as nature abhors a vacuum.

What is the best Vacuum that can be produced even with the most perfect machinery?—About thirteen pounds per square inch.

What is the function of the set-screw?—To hold any part of the machinery in its proper position after being adjusted.

What is friction?—Friction is the resistance offered by two substances to be rubbed or slided together.

What is the cause of abrasion?—Abrasion results from the want of sufficient lubrication, resulting in heating and rapid wasting of material.

Is perfect lubrication possible under all condi-

5 *

tions?—No; it is sometimes impossible to lubricate bearings in consequence of the extreme weight to which they are frequently subjected, and the speed at which they are run.

What is the meaning of the term Lubrication?—Oiling.

What is the meaning of the term Stroke when applied to a steam-engine?—A stroke is half a revolution.

How far does a Crank travel in order to make a stroke?—From one dead centre to the other.

When is the Crank at the dead centre?—When the centre of the crank-pin is parallel with the centre of the cylinder.

If an engine had a 24-inch stroke, how long would the crank have to be?—Twelve inches between the centre of the crank-shaft and the centre of the crank-pin.

Does the piston travel further on one-half of the stroke than on the other?—Yes; it travels further on the out-board than on the in-board half of the stroke, because the connecting-rod is shortening in the former case, while it is lengthening in the latter.

Is a piston in the centre of the cylinder when the crank is at half-stroke?—No.

Is it possible to design an engine in which the

piston would be in the centre of the cylinder when the crank is at half-stroke?—No.

Is there any loss of power in the employment of the crank?—No.

Is the power of the crank the same at all points of the stroke?—Yes; it is in proportion to the force expended, because, before the crank approaches the centre, the valve closes, and the rest of the work is performed by the momentum of the balance-wheel.

Is a Cut-off effected by the use of the Link capable of producing as satisfactory results as the Automatic arrangement?—No; because, as the cut-off is increased in the case of the link, the *lead* is also increased. All that can be said in favor of the *link* is that it is a convenient reversing gear.

How would you find the stroke of any engine?— By measuring the distance from the centre of the crank-shaft to the centre of the crank-pin.

How would you find the proper diameter of a governor-pulley for the engine-shaft?—I would multiply the revolutions of the governor by the diameter of the governor-pulley, and divide by the number of revolutions of the engine.

How do you find the number of feet the piston of a steam-engine travels per minute?—Multiply the distance it travels in inches for one stroke; multiply this product by the whole number of strokes, and divide by 12.

Into what two divisions may the cranks of all steam-engines be divided?—To single and double.

What is a Disk-crank, and what is the object of such cranks?—A disk-crank is a round plate with a crank-pin inserted at a distance suitable to meet the requirements of the stroke, and a counter-balance on the opposite side equal to the weight of the crank-pin and half the connecting-rod. They are generally used in high-speed engines, as they afford better facilities for balancing than either the single or double crank.

Why is an engine termed a throttling-engine?—Because the steam passes through the governor, and is throttled, cut or choked off by it, according to the circumstances of load and pressure.

Why is an engine termed an Automatic cut-off engine?—Because the valves for the admission and release of the steam are controlled by the governor outside of the steam-chest; whereas, in the case of the throttling-engine, the steam-valves give the same opening, whether the engine is doing much work or not, the supply being regulated by the governor-valve, while in the case of the automatic cut-off engine, the valves admit just sufficient steam to meet the requirements of load and pressure, and no more. The steam may be cut off at $\frac{1}{4}$, $\frac{1}{2}$, $\frac{5}{8}$, $\frac{3}{4}$, and $\frac{7}{8}$, as the case may be; but in the throttling-engine the steam is always cut off at the same point.

Can you give the names of the different cut-offs employed in steam-engines?—Positive, automatic, and variable.

Can you explain the difference between the three different arrangements?—In the positive, the cut-off is effected by what is known as lap on the valve; in the automatic it is controlled by the governor, where in the variable or adjustable the cut-off is regulated by a hand-wheel or screw outside of the steam-chest. This last arrangement is most generally used on steamships.

What is the object of a Double-crank?—Double-cranks may be used in the middle of a shaft, as in marine engines or stationary engines, which have a pillow-block on each side of the bed-plate. All locomotives had double-cranks at one time; they are now generally used on marine engines. Double-cranks are invariably made of wrought iron or steel.

How would you increase the speed of a steam-engine?—By enlarging the size of the pulley on the governor-shaft, because, if the governor runs slower, the engine will run faster.

Is there any power in the governor or fly-wheel of a steam-engine?—No; the governor performs the same function as the bridle does with a horse, while the fly-wheel only gives back the power it receives from the engine when it was put in motion.

What is the object of cutting off steam at a given

point in the Cylinder?—To get the benefits resulting from the expansion of the steam.

What is the meaning of the term Whole-stroke Engine?—An engine in which the valves are so arranged as to allow the steam to follow the piston ⅞ of the stroke.

Are Whole-stroke Engines in very general use? —No; though at one time all engines were whole-stroke, they are not built any more, except for some special purpose, on account of their wastefulness of steam and fuel.

What is an Eccentric? — A subterfuge for a crank.

Is an Eccentric a Cam?—No; an eccentric is a crank. The term cam has no definite meaning; it may have one, two, three, or four movements, but the movement of the eccentric is always the same.

How is the Throw of an Eccentric determined? —The throw of an eccentric must be proportioned to the travel of the valve or valves to which it is intended to give motion.

Would a small Crank perform the same function as an Eccentric?—Yes; precisely the same.

What are the advantages of the non-condensing over the condensing engine?—The first cost of the non-condensing is not more than half of that of the condensing engine; it is more simple, requires less skill, less management, can be run at almost any speed and under very high pressure, may be set up in any place where sufficient water can be obtained from which to evaporate the necessary volume of steam. Its disadvantage is its extreme wastefulness of fuel.

Will fresh and salt water unite in a steam-boiler? —No; as soon as salt water enters a boiler using fresh water, boiling or foaming is the result. The same effect is produced when fresh water is introduced into a boiler containing salt water.

How can you tell whether the boiler is foaming or not?—By the unsteady action of the water in the glass gauge.

When a boiler foams, can you tell with certainty how much water it contains?—No.

What are the most reliable means of finding the level of the water in a boiler when foaming?—Shut down the engine, cover the fire with fresh coal, shut the damper, and open the furnace door.

How may foaming arising from such circumstances be prevented?—By blowing out the salt water, and supplying the boiler with fresh water, and *vice versa*.

Is foaming in boilers dangerous?—Yes; as the water is lifted from the fire-plates, they become liable to be burned.

What is the meaning of the term Fuel?—Any

material employed to produce and sustain combustion.

Give the component parts of air in volume and weight.

Oxygen, 21 parts, Nitrogen, 79 parts, by volume; and by weight, Oxygen, 77 parts; Nitrogen, 23 parts.

Can air be condensed?—Yes; but not to meet any practical purpose, as, if air could be condensed like steam, we should not need the steam-engine.

Does air expand in volume by the application of heat?—Yes; but its expansive properties bear no proportion to that of steam.

Is there a class of engines called air-engines? —Yes; caloric engines, properly speaking, are air-engines.

Are caloric or air-engines capable of developing much power?—No; at one time it was anticipated by theorists that they would supersede the steam-engine, but, while the latter is increasing in numbers, the caloric engine is fast disappearing.

Does the caloric engine require a boiler similar to that of a steam-engine?—No; the cylinder is placed over the furnace, and the air drawn in by means of a pump, and on entering the hot cylinder it expands and increases in volume to a limited extent; the upper end of the cylinder of such engines is open to the pressure of the atmosphere.

What is the meaning of the term Combustion? —Burning.

What are the component parts of Anthracite and Bituminous coals?—Carbon and volatile matter. Anthracite coal consists almost entirely of pure carbon, while bituminous consists of carbon, hydrogen, oxygen, nitrogen, and other mineral compounds.

What is the meaning of the term Priming in steam-cylinders?—It means the passage of water from the boiler to the cylinder.

What is the cause of Priming in steam-cylinders? —It is generally caused by an insufficiency of steam-room in the boiler, or by the steam-pipe being too small to supply the cylinder, the effect of which is that when the valve opens the flow of the steam is so rapid that it carries the water with it.

What is the effect of Priming?—It causes a great waste of fuel, and involves a certain amount of danger, as the quantity of water carried over is liable to cause fracture of the cylinder or piston.

What is the best preventive against Priming? —Ample steam-room, a steam-pipe of sufficient area, and dry steam.

What is the meaning of the term saturated steam? —Steam which contains a certain amount of water in the shape of spray.

6

What is meant by the term Superheated steam? —Steam which has been dried by being exposed to extra heat after leaving the boiler.

Which has the more elastic force, saturated or superheated steam?—Superheated steam.

Can steam be generated by applying heat to the top of a vessel containing water?—No.

What designs of steam-engines are in most general use?—Vertical for marine and horizontal for stationary and locomotive purposes.

Do vertical engines possess any advantage over horizontal, and *vice versa?*—No; their employment is only a matter of convenience.

If the crank-pin or main-bearing should heat badly, what course would you pursue?—I would slack up on the key or nuts, and apply oil plentifully, and if the heating was extreme, I would first cool off with water.

Do you know of any substance that has a cooling effect on bearings?—Yes; plumbago, flour of sulphur, or quicksilver.

What is the cause of heating in Crank-pins or Pillow-block bearings?—Crank-pins and main-bearings heat from the following causes: grit or sand getting into the boxes, the shaft or pin not possessing sufficient area for the strain to which it is exposed, the load being so great as to force out or expel the lubrication, or the engine being out of line.

Are there any other causes except those above mentioned which produce heating in the different parts of steam-engines?—Yes; bad design, bad proportions, high speed, overwork, etc.

What is a propeller?—A mechanical device employed for propelling a vessel in water.

What principles are embodied in the design of the propeller?—The propeller is simply a screw similar to a nut on a bolt.

Has the propeller any advantages over the paddle-wheel as a means of propulsion?—No; in speed the paddle is superior to the screw in shallow water, while the propeller is superior to the paddle in deep water; but the paddle is more exposed to the oscillation of the wind or to the shots of an enemy in time of war.

What is the meaning of the term pitch of the propeller?—The term pitch of a propeller means the distance from the centre of one blade to the centre of the other.

What is the meaning of the term slip of the propeller?—The term slip means the distance the screw lacks in making one revolution in the water, compared with the distance it would have travelled if moving in a solid nut instead of in the water.

Is the paddle-wheel very generally employed at the present day?—No; it is fast giving place to the propeller, except for ferries, river, and lake service; but for ocean propulsion the screw has almost entirely superseded the paddle.

How long have you been on a marine engine, and in what capacity have you served?—Three years, as assistant engineer.

On what vessel were you employed, and between what ports did it ply?—The steamship Tallapoosa, between New York and New Orleans.

What kind of engines were employed on the ship—simple, compound, vertical, or incline?—Vertical compound engines.

What was the name of the captain in charge of the steamer?—L. S. Johnson.

What is an Indicator?—An instrument used for demonstrating the power which the engine is exerting, and showing the condition of the valves, piston, etc.

Can you draw an Indicator diagram and explain it?—Yes.

How would you proceed to obtain a marine license?—I would procure a blank form at the supervisor's office, fill it out, have it witnessed, have an affidavit before a United States commissioner, made by myself and my witnesses, certifying to the time during which I was on a steamship, tug, or ferry. I would then present myself for examination before the supervisor, and if I passed I would receive a license, which would allow me to run a vessel of a certain tonnage.

Can an engineer's license be revoked?—Yes; for incapacity, neglect of duty, or drunkenness.

How would you proceed to procure a certificate which would authorize you to take charge of a stationary engine?—I would present myself before the chief inspector or commissioner for examination; then, if I answered all the questions propounded to me correctly, I would be entitled to a certificate.

Do locomotive engineers require a license?—No.

How would you find the area of a circle?— Square the diameter and multiply by decimal ·7854.

How would you find the circumference of a circle?—Multiply the diameter by 3·1416.

What is the horse-power of a steam-engine?— 33,000 pounds raised one foot high in one minute.

How shall that 33,000 pounds be raised?—By belts or pulleys, or any other mechanical device that is most practicable and convenient.

How would you proceed to find the horse-power of an engine?—I would multiply the area of the piston by the average pressure, multiply this product by the number of feet the piston travels in

6 *   E

a minute, and divide by 33,000; the quotient will be the horse-power.

How would you find the area of the piston ?— I would square the diameter and multiply by the decimal ·7854, because the square is ·87 of the circle.

How would you find the horse-power of a steam-boiler ?—I would divide the heating surface by 16; the quotient would be the horse-power.

What amount of grate surface is recognized as the proper factor for a horse-power in steam-boilers?— One square foot for cylinder boilers, three-quarters for flue boilers, and one-half for tubular boilers.

What proportion should the safety-valve bear to the heating and grate surface ?—One-half square inch of safety-valve to one square foot of grate surface and sixteen square feet of heating surface.

How much water and how much coal does it require to develop a horse-power in a steam-engine? —About forty pounds of water and about four pounds of coal in ordinary engines; but the best class of automatic cut-off engines will produce the same result with a consumption of twenty-two pounds of water and two and a half pounds of coal.

ROPER'S

PRACTICAL

Handy-Books for Engineers.

———o○⦂⦿⦂○o———

ROPER'S ENGINEERS' HANDY-BOOK.

The most comprehensive and best illustrated book
ever published in this country on the Steam-Engine;
Stationary, Locomotive and Marine, and the Steam-
Engine Indicator. It contains nearly 300 Main Subjects; 1,316 Paragraphs, 876 Questions and Answers,
52 Suggestions and Instructions, 105 Rules, Formulæ,
and Examples, 149 Tables, 195 Illustrations, 31 Indicator Diagrams, and 167 Technical Terms; over 3,000
different subjects, with the questions most likely to be
asked when under examination, before being com-
missioned as an Engineer in the U. S. Navy or Revenue
Service, or licensed as an Engineer in the Mercantile
Marine Service. *Price, $3.50.*

ROPER'S HAND-BOOK OF LAND AND MARINE ENGINES.

Containing a description and illustrations of every
description of Land and Marine Engine in use at the
date of its publication, whether simple or compound,
horizontal, vertical, beam, steeple, direct-acting, back-
action, geared, oscillating, trunk, or rotary, with rules
for their care and management. *Price, $3.50.*

1

ROPER'S HAND-BOOK OF MODERN STEAM FIRE-ENGINES.

The only book of the kind ever published in this country. It contains descriptions and illustrations of all the best types of Steam Fire-Engines, and Fire-Pumps, Injectors, Pulsometers, Inspirators, Hydraulic Rams, etc.; and treats more extensively on Hydraulics than any other book in the market. *Price, $3.50.*

ROPER'S HAND-BOOK OF THE LOCOMOTIVE.

One of the most valuable treatises ever written on the subject, as it is so plain and practical that any Engineer or Fireman that can read can easily understand it. It is fully illustrated, and contains a description of the most improved types of Locomotives in use. *Price, $2.50.*

ROPER'S INSTRUCTIONS AND SUGGESTIONS FOR ENGINEERS AND FIREMEN.

This little book is made up of a series of suggestions and instructions, the result of recent experiments and the best modern practice in the care of steam-engines and boilers. It is brimful of just such information as persons of limited education having charge of steam machinery need. It is written in plain, practical language, devoid of theories or mathematical formulæ. *Price, $2.00.*

ROPER'S USE AND ABUSE OF THE STEAM-BOILER.

Containing Illustrations and Descriptions of all classes of Steam-Boilers in use at the present day; Plain Cylinder, Flue, Double-deck, Tubular, Tubulous, Patent, etc., with instructions how to set up, fire, and manage the same. *Price, $2.00.*

ROPER'S CATECHISM OF HIGH-PRESSURE STEAM-ENGINES.

Written in the form of Question and Answer, for the use of Engineers of limited education and experience. It contains a fund of valuable information for Engineers, expressed in plain, simple language. *Price, $2.00.*

ROPER'S QUESTIONS AND ANSWERS FOR ENGINEERS

Contains all the Questions that an Engineer will be asked when undergoing an examination for the purpose of procuring a license, with the Answers to the same, couched in language so plain that any Engineer or Fireman can in a short time commit them to memory. *Price, $3.00.*

ROPER'S SIMPLE PROCESS

For estimating the horse-power of Steam-Engines, from indicator diagrams, or the work an engine is performing at the time the diagram was taken. One of the most important devices ever employed in connection with the Steam-Engine. *Price, 50 cents.*

ROPER'S CARE AND MANAGEMENT OF THE STEAM-BOILER.

One of the most practical works ever published on this subject, as it embraces the following subjects: Care and Management of Steam-Boilers, Horse-power of Steam-Boilers, Repairing Steam-Boilers, Incrustation in Steam-Boilers, Steam-Boiler Explosions. Testing Steam-Boilers, Externally and Internally Fired Steam-Boilers, Design of Steam-Boilers, Steam-Boiler Materials, Mud-Drums, Steam-Domes, Cleaning Steam-Boilers, Different Types of Steam-Boilers, Feed-Water Heaters, Fuel, Chimneys (area and height), Draught, Smoke, Instructions for Firing, Comparative Efficiency of Different Types of Steam-Boilers, with a great amount of other information of immense value to owners of Steam-Boilers, Engineers, and Firemen, expressed in plain, practical language. *Price, $2.00.*

ROPER'S IMPROVEMENTS, IN STEAM-ENGINES.

This is the most comprehensive, elaborate, and best illustrated work ever published on the steam-engine, as it contains illustrations and descriptions of all modern steam-engines of every class—stationary, locomotive. marine, hoisting, winding, pumping, traction, fire, portable, rotary, trochilic, Herishof, torpedo, simple, compound, trunk, oscillating, direct-acting, back-action, steeple, inclined, upright. vertical, Corliss and Cornish, with a description of the latest improvement in the same.   In preparation.

E. CLAXTON & CO., Publishers,
930 Market St., Philadelphia, Pa.

www.ingramcontent.com/pod-product-compliance
Lightning Source LLC
Chambersburg PA
CBHW022008190326
41519CB00010B/1439